russian aircraft in action

# Tupolev
# Tu-95/-142

## Yefim Gordon

**Tupolev Tu-95/Tu-142**
© 2003 Yefim Gordon
ISBN 1 932525 00 9

Published by IP Media, Inc.
350 Third Avenue, PMB 368
New York, NY 10010, USA
Tel: 718 305 3355   Fax: 718 305 3356
E-mail: ipmediainc@rcn.com

© 2003 IP Media, Inc.

Design concept and layout
by Polygon-Press Ltd. (Moscow, Russia)
E-mail: info@polygonpress.com

Polygon-Press is an affiliate of IP Media

Printed in Slovakia

All rights reserved. No part of this publication may be reproduced, stored in a retrieval system, transmitted in any form or by any means, electronic, mechanical or photo-copied, recorded or otherwise, without the written permission of the publishers.

# Contents

| | |
|---|---|
| Tu-95/Tu-142 – a Brief History | 3 |
| Photo Gallery | 6 |
| Tu-95RTs Line drawings | 76 |
| Tu-95 & TU-142MR Colour Profiles | 78 |

This book is illustrated with photos by Yefim Gordon, Andrey Grischchenko and Sergey Skrynnikov, as well as photos from the archives of Yefim Gordon, Vladimir Rigmant, the Tupolev Joint-Stock Company, Nigel Eastaway of the Russian Aviation Research Trust, the Royal Air Force and the Royal Swedish Air Force.

Colour artwork by Andrey Yurgenson

The author wishes to thank Sergey and Dmitriy Komissarov for their assistance with the English introductory text and photo captions.

# Tupolev Tu-95/Tu-142

The story of the Tu-95 cannot be properly understood without mentioning some events of the preceding period. In 1943 the Soviet Government, being well informed through intelligence channels of the US nuclear weapon programme, launched a similar programme of its own. The programme's thrust was to create and test the first Soviet atomic bomb within a short time interval with the objective of countering the perceived post-war threat from the West. A problem quickly surfaced: how to deliver that weapon to the target.

The heavy bomber arm of the Soviet Air Force (DA – **Dahl'**nyaya avi**ah**tsiya, Long-Range Aviation), which came into existence in 1942, had not been created initially with intercontinental strikes in mind. This issue arose in conjunction with the nuclear weapons programme. The problem was that none of the bombers in service with the Soviet Air Force at the end of the war had the required range or stood a chance against contemporary fighter forces. To tackle the issue, in 1944 the OKB-156 experimental design bureau (**o**pytno-**kon**strook**t**orskoye byuro) led by Andrey Nikolayevich Tupolev was tasked with designing a heavy bomber possessing intercontinental range. The OKB came up with a number of projects but soon ran into development problems. This prompted the Sovier leaders to choose a "quick fix". Tupolev was tasked with producing an exact copy of the Boeing B-29. This decision was undoubtedly influenced by the availability of three B-29s which had force-landed in the Soviet Far East.

The result of this copying, the Tu-4 bomber, marked a turning point for the Soviet aviation industry. Gaining access to B-29 technology enabled it to make a truly giant leap which laid the foundation of all future heavy bomber development in the USSR. Coupled with the creation of Soviet nuclear weapons, this had a profound influence on the relationship between the Soviet Union and the Western world.

However, the prime target – the United States – was beyond the range of the Tu-4, especially when it was carrying a maximum payload such as an atomic bomb. Hence, once the Tu-4 had entered production, Tupolev began looking for ways to obtain intercontinental range. These studies resulted in the Tu-80 and Tu-85 bombers which entered test in December 1949 and January 1951 respectively. The latter design became the true forerunner of the Tu-95. The Tu-85 met its range and payload targets, and production plans were already in hand when it became obvious that a piston-engined bomber would have few chances when opposed by the jet-powered interceptors under development in the USA. Assessing bomber development in the West, the Soviet military quickly concluded that advanced heavy bombers should be powered by either turboprop or turbojet engines.

In 1950 Vladimir M. Myasischchev offered to create a jet-powered strategic bomber with a top speed of 950 km/h (590 mph) and a range in excess of 13,000 km (8,070 miles). The project was given a high degree of priority, eventually emerging as the M-4. Tupolev knew about the development of the M-4; unsurprisingly, he wanted to compete with Myasischev for the strategic fast bomber order from the Soviet Air Force. In the spring of 1948 OKB-156 began preliminary design (PD) work. A study of captured German research into large swept-wing aircraft was undertaken by the Central Aero- & Hydrodynamics Institute (TsAGI) which declared swept wings to be the best choice for Tupolev's new strategic bomber.

A take-off weight of some 150 tons (330,690 lb) and wings swept back 35° at quarter-chord, with an aspect ratio of 9, would be required to achieve the requisite range and speed-at-altitude coupled with a reasonable bomb load. The various powerplant options were quickly narrowed to either turbojet or turboprop engines. Eventually, the PD studies showed that a turboprop-powered aircraft was the best option for reaching the required range of over 13,000 km (8,073 miles), with a top speed of about 800 km/h (497 mph). In comparison, a turbojet-powered version offered a range of only 10,000 km (6,210 miles) and a top speed of 900 km/h (559 mph). The high priority placed on range led Tupolev to go for turboprops.

The VVS, aware of the progress made by Myasischchev with the M-4 jet bomber, did not agree with Tupolev's decision. Tupolev had to defend his choice at a meeting with the Soviet leader Iosif V. Stalin who elected to continue work on the M-4, approving initial development funding for the Tupolev strategic bomber at the same time. The aircraft was henceforth referred to as the "**95**".

OKB-276 headed by Nikolay Dmitriyevich Kuznetsov had made some progress in turboprop engine development, making use of captured German technology. Evolved from the Junkers Jumo 022, the TV-2 engine passed bench tests in October 1950; an improved version delivered 4,620 ehp. The TV-10 and TV-12 engines followed, producing 10,000 and 12,000 ehp respectively. At the time, these were by far the most powerful turboprops in the world. However, development delays led Tupolev, upon consultation with Kuznetsov, to resort to a temporary solution involving the use of the smaller TV-2F in a paired configuration, the two engines being coupled to a common gearbox. The resulting powerplant was designated 2TV-2F. It was decided that contra-rotating propellers would be best for the new engine, avoiding problems associated with an unacceptably big propeller diameter.

Final engineering design of the "95" started on 15th July 1951. In accordance with the Soviet Air Force's requirement and a Council of Ministers directive, the aircraft was to have a range of up to 15,000 km (9,315 miles) with a payload and an absolute maximum range of 17,000-18,000km (10,560-11,180 miles). Cruising speed was specified as 750-820 km/h (466-509 mph) and maximum speed as 920-950 km/h (571-590 mph). Service ceiling was expected to be 14,000 m (45,920 ft). These performance estimates were somewhat revised later and included a cruising speed of 750 to 800 km/h (466-497 mph) at 10,000-14,000 m (32,800-45,920 ft). Maximum range was now estimated as 14,500-17,500km (9,000-10,870 miles).

A comparison of the versions powered by 2TV-2Fs and TV-12s prompted the decision to proceed with the 2TV-2F pending the availability of the more powerful TV-12, selecting the latter engine as the standard powerplant for the production bomber.

The maximum bomb load was 15,000 kg (33,070 lb) and the normal bomb load 5,000 kg (11,020 lb). Conventional bombs, torpedoes, mines and other weapons could be carried; the bomb bay was heated, enabling the carriage of nuclear munitions. The aircraft had a crew of eight.

The control system proved to be a major challenge. TsAGI insisted that irreversible hydraulic actuators be used; Tupolev argued that the actuators had a poor reliability record and that the old but dependable mechanical systems then in use were sufficient. Reluctantly, TsAGI capitulated.

The first prototype (known as the "95-1") powered by 2TV-2F engines was completed in the autumn of 1952. The first flight took place on 12th November 1952; the crew was captained by test pilot A. D. Perelyot.

Tragically, on 11th May 1953 the prototype crashed on its 17th flight after suffering an engine fire and loss of control. Four of the eleven crew members, including A. D. Perelyot, did not bail out and died in the crash, trying to the last to save the prototype. Later, Perelyot was posthumously awarded the title of Hero of the Soviet Union. The cause of the accident was traced to the No. 3 engine's reduction gearbox which had failed because of metal fatigue and improper material choice. Appropriate steps were taken by Tupolev and Kuznetsov to ensure that the TV-12 turboprop envisaged for the second prototype be thoroughly tested on the ground and on a flying tesbed before being installed on the actual aircraft.

The airframe of the secondprototype (the «95-2») was completed in November 1952 but it was not until December 1954 that the TV-12 engines (by then redesignated NK-12) were finally installed. This delay was due partly to Kuznetsov's cautious approach; he wanted to make sure there would be no replay of the May 1953 tragedy. Meanwhile, aircraft factory No. 18 in in Kuybyshev (now Samara) initiated production of the type under the in-house code "*izdeliye* (product) V".

On 16th February 1955 the second prototype finally flew with M. A. Nyukhtikov and I. M. Sookhomlin at the controls. The tests were completed on 8th January 1956 without major incident.

During the summer of 1955 the "95" was demonstrated publicly for the first time at the annual Aviation Day flypast. The big turboprop bomber made a strong impression on Western aviation experts. Within a short time the "95" received the NATO reporting name *Bear*; the original bomber became the *Bear-A* when other versions were identified. At first it was erroneously ascribed to the Il'yushin design bureau and referred to as the IL-38 (which is actually a totally different aircraft – an anti-submarine warfare derivative of the IL-18 four-turboprop airliner). Later, when its Tupolev identity was established, for many years the bomber was called Tu-20 in the West.

In the meantime, the trials continued unabated. A maximum-range test flight made in the autumn of 1955 covered a distance of 13,900 km (8,630 miles), which was 1,100 km (683 miles) less than the specified range but still sufficient to reach the primary target – the North American continent.

The first production-standard aircraft bearing the service designation **Tu-95** entered test in August 1955. They differed from the prototypes in having a 2-m (6.5-ft) fuselage stretch, a complete systems fit and a 5% increase in empty weight. The stretch was needed to increase fuel capacity, as the NK-12 proved to be somewhat thirstier than anticipated.

The performance figures obtained at this stage were considerably lower than anticipated. Hence in August 1956 – February 1957 the Tupolev OKB converted the first production aircraft (c/n 5800101) into the prototype of the **Tu-95M** (*modifitseerovannyy* – modified) or *izdeliye* VM. The new version featured 15,000-ehp NK-12M turboprops with a lower fuel consumption. The MTOW grew from 172,000 kg (379,188 lb) to 182,000 kg (401,234 lb) and the fuel load from 80,730 kg (177,980 lb) to 89,530 kg (197,380 lb). The Tu-95M completed trials in 1958, showing a range of 13,200 m (8,200 miles), a top speed of 902 km/h (560 mph) and a cruising speed of 720-750km/h (447-466 mph). 18 such aircraft were delivered to the Soviet Air Force; by comparison, production of the Tu-95 *sans suffixe* totalled 31. These bombers became the first Tu-95s to enter service.

Later, many Bear-As received a mid-life update involving installation of a new Rubin-1D (Ruby) bomb-aiming radar replacing the Rubidiy-MM, a new ADNS-4 long-range radio navigation system, an RSBN-2SV Svod (Dome) short-range radio navigation system etc.

A version of the Tu-95M designated Tu-95A (**ah**tomnyy – nuclear-capable) was developed specifically for delivering nuclear weapons. Differences included a heated and air-conditioned bomb bay and reflective white undersurfaces for protection from the heat generated by nuclear explosions.

By the late 1980s all surviving Bear-As had been converted to Tu-95U trainers (oo**cheb**nyy – training, used attributively). The weapon systems were removed and the bomb bay faired over. The Tu-95Us wore a distinctive red stripe around the aft fuselage. A single production Tu-95 was converted into an electronic countermeasures (active jamming) platform in the late 1960s.

To meet an Air Force requirement the OKB developed a reconnaissance version designated **Tu-95MR** ([*samolyot*-] *razvedchik*, reconnaissance aircraft), aka izdeliye VR or Bear-E. The Tu-95MR was capable of both photo reconnaissance (PHOTINT) and electronic reconnaissance (ELINT), with dielectric antenna blisters on the aft fuselage and a battery of cameras in the bomb bay. The aircraft was equipped with a probe-and-drogue in-flight refuelling (IFR) system. Four Tu-95Ms (c/ns 7800410, 7800501, 7800502 and 7800506) were upgraded to this configuration.

Back in 1952 the Tupolev OKB initiated development of a high-altitude version designated **Tu-96**. Designed to operate at up to 16,000-17,000 m (52,480-55,760 ft), the aircraft had increased-area wings (as later fitted to the Tu-114 airliner derivative of the Tu-95) and was to be fitted with high-altitude Kuznetsov TV-16 turboprops then under development, which produced 12,500 ehp up 14,000 m (45,920 ft). Range under varying conditions was to be 9,000-18,000 km (5,590-11,180 miles). The aircraft was to have a top speed of 900-950 km/h (559-590 mph) at 8,000-9,000 m (26,240-29,520 ft). Work on the experimental TV-16 engine did not progress as rapidly as expected; the engines were not available in time for the initial flight tests, forcing Tupolev to temporarily install NK-12s on the sole prototype. Meanwhile, however, the VVS reconsidered its approach to high-altitude strategic bombers which were found to be too vulnerable to the latest Western fighters and surface-to-air missiles, and the Tu-96 was abandoned.

To counter the rapid development of Western air defence systems the OKB brought out a missile strike version. Designated **Tu-95K**, it was armed with a single Kh-20 (NATO AS-3 *Kangaroo*) air-to-surface missile – a large stand-off weapon developed by the Mikoyan design bureau . The K suffix stood for the *Kometa* (Comet) code name of the Soviet anti-shipping missile programme; later on, it was deciphered as **kom**pleks [*vo'oroozheniya*] – weapons system. The Tu-95K featured a YaD target illumination radar (NATO *Crown Drum*) in the nose, supplanting the glazed nose of the Bear-A; the twin-antenna array of the YaD (one antenna for search and another for missile guidance) resulted in a distinctive "duck bill" nose profile. The turbojet-powered Kh-20 was carried semi-recessed in the fuselage and extended into the slipstream prior to launch.

The two Tu-95K prototypes converted from *Bear-As* (c/ns 5800001 and 6800404) were tested in 1956. Production began in March 1958 as *izdeliye* VK and the type was officially included into the Soviet Air Force inventory on 9th September 1960. The NATO code name was *Bear-B*. In the late 1980s the remaining Tu-95Ks were converted into **Tu-95KU** (*izdeliye* VKU) trainers. A longer-range version featuring IFR capability and electronic countermeasures (ECM) gear entered production in 1962 as the **Tu-95KM** (*izdeliye* VKM or *Bear-C*); the missile it carried was upgraded accordingly as the Kh-20M. The total number of Tu-95K/Tu-95KM missile carriers completed was 71.

Later, a number of Bear-Cs was upgraded to **Tu-95K-22** (*izdeliye* VK-22/*Bear-G*) standard by integrating Kh-22 (AS-4 *Kitchen*) stand-off missiles. The reconfigured aircraft featured a BD-45F centreline rack for the Kh-22, augmented by two BD-45K pylons under the wing roots for carrying one Kh-22 under each wing. A new PNA-B (NATO *Puff Ball*) radar was fitted and the tail gunner's station gave place to an ECM fairing called UKhO (oonifi-**tsee**rovannyy khvosto**voy** otsek – standardised tail bay). The first flight took place on 30th October 1975. Several aircraft were equipped with RR8311-100 air sampling pods on underwing pylons for radiation reconnaissance duties; this system was developed specifically to explore the results of above-ground nuclear testing, primarily in China.

In an attempt to integrate the KSR-5 (AS-6 *Kingfish*) air-to-surface missile on the *Bear* the Tupolev OKB converted a single Tu-95M (c/n 8800601) into the **Tu-95M-5** development aircraft in October 1976. The trials continued through May 1977 but were terminated when the Tu-95M-5 showed no significant advantages over other aircraft with this missile system.

In 1959 a single Tu-95 (c/n 5800302) was extensively modified for testing the Soviet Union's first hydrogen bomb – the 50-megaton *Vanya*. Accordingly the converted aircraft was designated **Tu-95V (Tu-95-202)**. For political reasons the first live drop of a Soviet H-bomb was delayed until 1961 (in 1959 the Soviet Premier Nikita S. Khruschchov was due to visit the USA, and it was inappropriate for such a weapon to be tested while the Soviet leader was on US soil). The test took place at the Novaya Zemlya test range on 30th October 1961, and the blast was devastating; suffice it to say that the shock wave ran around the globe three times!

In 1963 the OKB brought out an over-the-horizon (OTH)/maritime reconnaissance targeting reconnaissance version. Designated **Tu-95RTs** (*razvedchik-tseleookazahtel'* – recce/target designator aircraft) or *izdeliye* VTs/*Bear-D*, it was part of the world's first maritime reconnaissance/strike complex, providing mid-course guidance for the P-6 anti-shipping missiles launched by Soviet submarines. To this end the Tu-95RTs was equipped with the Oospekh-1A (Success-1A) 360° search radar in a huge teardrop radome and an Arfa (Harp) data link system for relaying guidance signals from the sub to the missile, with antennas at the tips of the horizontal tail and under the nose.The Tu-95RTs entered production in 1963, and a total of 53 was built; initial operational capability with the Soviet Navy was attained in August 1964. The type remained operational until the late 1980s.

The **Tu-142** long-range anti-submarine warfare aircraft (*izdeliye* VP/*Bear-F*) represented a major "rejuvenation" of the Tu-95. The redesign was extensive enough to warrant an entirely new designation. The aircraft had new wings of greater area and a different airfoil section which incorporated integral fuel tanks replacing the bag-type tanks of the Tu-95 were replaced by in the wing torsion box; irreversible hydraulic actuators were installed at last to reduce pilot workload. The Tu-142 had a search and targeting system (STS) built around a Berkoot (Golden eagle) 360° search radar mounted ventrally at near mid-fuselage, a sophisticated navigation system and an ASW weapons system. The latter included torpedoes, mines and depth charges.

The first prototype (c/n 4200) took to the air on 18th July 1968. Deliveries to the Naval Air Arm began in 1970, the aircraft initially serving with the North Fleet. The specification for the Tu-142 contained an unorthodox clause: the aircraft had to be capable of operating from semi-prepared airfields. Hence the initial production version had 12-wheel main gear bogies which earned it the nickname *Sorokonozhka* (Centipede). However, the requirement proved superfluous and the 12-wheel bogies soon gave way to lighter four-wheel bogies in slimmer fairings.

In the mid-1970s Tu-142 production switched from Kuybyshev to the Taganrog machinery plant No. 86. This coincided with the appearance of a new version designated **Tu-142M** (*izdeliye* VPM/*Bear-F Mod 2*). This aircraft featured a 0.3-m (1-ft) forward fuselage stretch and a redesigned flight deck. The NATO reporting name was Bear-F Mod 2.

The next version, the **Tu-142MK** (*izdeliye* VPMK/*Bear-F Mod 3*), featured a new STS designed around the Korshoon (kite, the bird) search radar, hence the K suffix. The Tu-142MK was the first version to be equipped with a magnetic anomaly detector with a characteristic MAD "stinger" atop the fin. Development was beset by teething troubles and it was a while before the aircraft, which achieved IOC in 1980, became fully operational. Eight *Bear-F Mod 3s* were delivered to the Indian Navy in **Tu-142MK-E** export configuration in the 1990s. The final version which entered flight test in 1985 and became fully operational in 1993 was the **Tu-142MZ** (*izdeliye* VPMZ/*Bear-F Mod 4*) fitted with an upgraded ASW suite, improved NK-12MP engines and a TA-12 auxiliary power unit. The last Tu-142 left the plant in 1994, putting an end to the long production of the *Bear* family.

A version designated **Tu-142MR** (*izdeliye* VPMR/*Bear-J*) was developed by the Beriyev OKB for the Soviet Navy as a communications relay aircraft between submerged nuclear missile submarines and land-based or airborne command posts in the event of a nuclear attack. It was equipped with a trailing wire aerial (TWA) for very low frequency (VLF) communications.

The one-off **Tu-142MRTs** maritime reconnaissance aircraft based on the Tu-95MS and Tu-142M was tested in the late 1980s/early 1990s. As the designation reveals, this was a prospective replacement for the Tu-95RTs. However, the remained a due to a change in priorities by the Soviet Navy. Unfortunately no details are known.

An experimental version designated **Tu-95M-55** took to the air on 31st July 1978. It paved the way for the **Tu-95MS** cruise missile carrier (*izdeliye* VP-021/*Bear-H*) based on the Tu-142MK airframe which entered flight test in September 1979. The *Bear-H* came in two configurations designated **Tu-95MS-6** and **Tu-95MS-16**; the former was armed with six Raduga Kh-55 (AS-15 *Kent*) cruise missiles on an MKU-6-5 rotary launcher, while the other version carried an additional ten missiles on underwing pylons (triple racks under the wing roots and double rack between the engines). The nose accommodated a new Obzor (Perspective) target illumination/guidance radar in a "duck bill" radome reminiscent of the Tu-95K/KM but rather neater. The aircraft was powered by NK-12MP turboprops.

Production began in Taganrog in 1981 and was transferred to Kuybyshev in 1983, totalling 31 Tu-95MS-6s and 57 Tu-95MS-16s. The underwing weapons pylons were later removed from the latter version under the terms of the START-1 treaty. Together with the Tu-160, the Tu-95MS remains the mainstay of Russian Strategic Aviation today.

A single production Tu-95MS was upgraded to a new configuration designated **Tu-95MA** in early 1993. Two large pylons were fitted under the wing roots for carrying large advanced missiles of undisclosed designation.

The more exotic variants of the Bear included the **Tu-95N (Tu-95RS)** "mothership" for the Tsybin RSR supersonic strategic reconnaissance aircraft converted from the first production Tu-95 ("4807 Black", c/n 5800101) in 1959; the **"mothership" for Mikoyan's "105.11" fighter aerospaceplane** converted from a Tu-95KM (c/n 63M52607) in 1977; the **Tu-95 Vostok spacecraft locator aircraft** of late 1960 designed to find the re-entry capsules of Vostok manned spacecraft; and the **Tu-116** long-range VIP aircraft derived from the *Bear-A*, of which two were built.

In 1957-58 the second prototype Tu-95 was converted into the **Tu-95LL** testbed for the Flight Research Institute in Zhukovskiy. It served for testing new powerful jet engines, including the NK-6, NK-144 and NK-22 afterburning turbofans. The development engine was mounted in a special nacelle suspended from a hydraulically-powered trapeze in the bomb bay. The nacelle was lowered clear of the fuselage before engine starting. This role later passed to two *Bear-Fs* – the Tu-142 prototype (c/n 4200) and the Tu-142MK prototype (c/n 4243) – converted into **Tu-142LL** engine testbeds. Stripped of all ASW equipment and armament, they were used to test the NK-25, NK-32 and Kolesov RD36-51A afterburning turbofans.

The weirdest version of all, however, was the **Tu-95LAL** research aircraft converted from a production Tu-95M ("51 Red", c/n 7800408) in 1961 to explore the feasibility of nuclear propulsion systems for aircraft. It featured a VVRL-100 nuclear reactor in the bomb bay and a radiation barrier between the crew and the reactor.

The type's operational history began in April 1956. During the Cold War the mighty Tu-95/Tu-142 was one of the symbols of the tell-tale «Soviet threat», maintaining a constant presence in various corners of the world. The change in the Soviet political situation in the late 1980s and the subsequent collapse of the Soviet Union had an adverse effect on the type's operations. Yet the *Bear* is far from dead – it remains an important element of Russia's air arm which, plagued as it was by fuel shortages and reorganizations, managed to keep up operational levels. In 2000 the Russian Air Force obtained several Tu-9MS-6 from the Ukraine, thus saving them from the breaker's torch. Boasting the same longevity as its American counterpart (the B-52), the venerable Tu-95 has a chance of soldiering on well into the 21st century.

◄ The first prototype of Tupolev's «aircraft 95» strategic bomber (known inside OKB-156 as the «95-1») was powered by Kuznetsov 2TV-2F coupled turboprops driving eight-bladed contrra-rotating propellers via common reduction gear. It was the failure of this gearbox and the resulting engine fire which caused the first prototype to crash fatally on 17th May 1953, bringing about a change of engine type to the Kuznetsov NK-12 from the second prototype onwards.

This desktop model shows how the ill-starred «95-1» looked; unfortunately no photographs of the actual aircraft have survived. The model illustrates the widened lower portions of the engine nacelles where the two TV-2F turbines making up each coupled engine were housed, breathing through a common full-width air intake. The slender fuselage, high aspect ratio wings and the main landing gear fairings characteristic of Tupolev aircraft are also evident. ▼

▲
The first production Tu-95 «sans suffixe» (construction number 5800101 – i.e., year of manufacture 1955, Kuibyshev aircraft factory No. 18, Batch 001, first aircraft in the batch) was serialled «6 Black». At the factory it bore the in-house product code «izdeliye (product) V». This view illustrates the cylindrical nacelles of the NK-12 engines and the extensively glazed navigator's station of the initial production version (NATO reporting name Bear-A). The tall landing gear was a necessity to provide adequate blade tip clearance; note the angle of the nose gear strut.

A three-quarters rear view of Tu-95 c/n 5800101. The different shades of metal on various skin panels gave the aircraft a rather patchwork appearance. This view shows the tail gunner's station, the lateral sighting blisters for the gunner operating the dorsal and ventral barbettes and the retractable tail bumper equipped with twin wheels.
▶

No, the number 4807 on the nose and tail of Tu-95 «46 Black» is not the construction number but a meaningless number intended to lead would-be spies astray (this is actually the same first production aircraft, c/n 5800101). The aircraft was undergoing tests at the time, as revealed by the photo calibration markings on the fuselage. It was later converted into the Tu-95N «mother ship» for the Tsybin RSR high-speed reconnaissance aircraft.

This Tu-95M serialled «44 Black» was photographed when passing over Red Square during a military parade in Moscow in 1956. Note the exhaust stains aft of the bifurcated engine jetpipes; the Bear's main gear fairings were usually black with soot.

A trio of Tu-95Ms in Vee formation puts on a show of force during a military parade in Moscow in 1956.

This fine air-to-air study of a Tu-95 cruising above the clouds was taken by a shadowing NATO fighter. The undersurfaces and rudder painted gloss white identify this aircraft as a nuclear-capable Tu-95MA.

▲
A Tu-95M with the tactical code «42 Red» writ large on the nose takes part in another military parade in 1956 (possibly on May Day). The escorting Mikoyan/Gurevich MiG-17 «sans suffixe» (Fresco-A) fighters provide scale, giving an idea of the bomber's impressive size.

A Tu-95M flies over the Russian countryside. This view taken from a sister aircraft illustrates the sleek lines of this impressive aircraft.

◀

▲ A Tu-95A coded «44 Red» taxies at a Soviet Long-Range Aviation (o.e., strategic bomber arm) base. From the 1960s onwards the tactical code was carried in much smaller digits on the nosewheel well doors and the tail.

A grey-painted US Air Force McDonnell Douglas F-4E Phantom II maintains a close interest in a Tu-95A flying over international waters during a routine training mission. ▶

«73 Black», the sole prototype Tu-96 – a rewinged Bear-A featuring increased-area wings identical to those of the Tu-114 Rossiya (Russia) long-haul airliner which in turn was derived from the Tu-95. The intended TV-16 turboprops were unavailable and the aircraft was tested with stock NK-12 engines throughout; note the photo calibration markings. The number 5836 is again a red herring, not the c/n.

Front view of the second prototype of the Tu-95K strategic missile strike aircraft (izdeliye VK; NATO reporting name Bear-B). Instead of a glazed navigator's station the fuselage nose housed a twin-antenna YaD target illumination/ missile guidance radar. The target illumination channel antenna was wider than the fuselage, resulting in a characteristic «duck bill» radome.

▲
Another view of the second prototype Tu-95K (c/n 6800404) converted from a standard Bear-A which left the factory in 1956. This view shows the Bereznyak K-20 cruise missile (NATO code name AS-3 Kangaroo) semi-recessed in the fuselage, with a retractable fairing over the missile's air intake to keep the turbojet engine from windmilling in flight.

The uncoded second prototype Tu-95K featured a telemetry antenna array instead of the twin AM-23 cannons in the tail turret and a cinO camera in a teardrop-shaped fairing under the starboard wingtip to capture the missile's separation sequence.
▼

◀
To extend its combat radius the Tu-95K was equipped with a probe-and-drogue in-flight refuelling system in 1961, the resulting long-range version being designated Tu-95KM Bear-C. This is the forward fuselage of an updated Tu-95KM manufactured in 1960 (c/n 60802207); note the fuel line conduit running from the nose-mounted refuelling probe to the centre fuselage fuel tanks. The aircraft carries a K-20 missile.

New versions appeared as new weapons were integrated. This aircraft coded «91 Red» is a Tu-95K-22 upgrade (NATO reporting name Bear-G). The large Kh-22 (AS-4 Kitchen) cruise missiles carried on pylons under the inner wings are just discernible in this view. Note also the IFR probe, the nose «thimble» radome associated with missile guidance, the elongated electronic countermeasures (ECM) fairings on the rear fuselage sides and the ogival fairing (designated UKhO) with an SPS-153 Rezeda (Mignonette) jammer supplanting the tail gunner's station.

▼

In October 1976 a production Tu-95M (c/n 8800601) was converted into the sole Tu-95M-5 (izdeliye VM-5) prototype. The aircraft was armed with two KSR-5 missiles on underwing pylons, a modified Rubin-1KV radar and an SPS-153 jammer in a UKhO tail fairing. Seen here at Zhukovskiy in the winter of 1976, the aircraft underwent tests but the programme was terminated in May 1977.

▲◀ Towards the end of their flying careers the few Tu-95Ms were transferred to the Long-Range Aviation's 43rd TsBP i PLS (Combat and Conversion Training Centre) at Dyaghilevo airbase near Ryazan' in central Russia. Here an example coded «67 Red» is seen at Dyaghilevo AB in 1991. The aircraft is still intact, but the red band around the rear fuselage signifies it has been disarmed in compliance with arms reduction treaties, becoming a Tu-95MU trainer. Note the orange-painted ladder for access to the forward pressure cabin via the nosewheel well.

▲ Close-up of the forward fuselage of Tu-95MU «67 Red» at Dyaghilevo AB. The 'Ms were the last of the original «glass-nosed» Bear bombers to remain in service.

A Tu-95RTs over-the-horizon targeting/ maritime reconnaissance aircraft (izdeliye VTs, NATO reporting name Bear-D) in high-altitude cruise. This view emphasises the Bear's large wing area.

▼

With its mighty AV-60 contraprops gone, this well-worn Tu-95MU at Dyaghilevo appears destined never to fly again. It is coded «71 Red» at the 43rd Combat and Conversion Training Centre, but the previous tactical code «57 Blue» with which it was operated by another regiment is showing through the white paint on the nose gear door.
◀

Intended to designate targets for missiles launched by Soviet submarines in a war scenario, in peacetime the Tu-95RTs OTH aircraft had little to do but to shadow the movements of NATO navies – and were, in turn, shadowed by NATO fighters attempting to ward off the unwelcome guests. Such encounters over international waters were frequent in the 1970s.

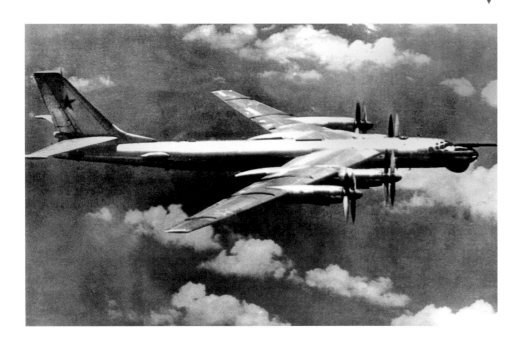

▲
A North Fleet Air Arm/392nd ODRAP (Independent Long-Range Reconnaissance Regiment) Tu-95RTs coded «20 Red» parked at a snow-covered dispersal area at Kipelovo AB near Vologda. These photos show the Bear-D's main identification features – the huge teardrop radome under the centre fuselage housing the Oospekh (Success) 360° search radar and the deep chin fairing associated with the submarine-to-missile command link system.
▼

Pacific Fleet Air Arm Tu-95RTs «38 Red» parked at Khorol' airbase in the Soviet Far East sports an «Excellent aircraft» badge (a stylised aircraft silhouette in a «Quality SIgn» pentagon). This badge was awarded for keeping an aircraft in perfect technical condition and placed a high degree of responsibility on the ground personnel maintaining the aircraft.

Another air-to-air of a Tu-95RTs cruising over thick overcast.

Like the Tu-95KM Bear-C and the Tu-95K-22 Bear-G, the Tu-95RTs featured large lateral ECM blisters. This view also shows the cigar-shaped fairings of the Lira (Lyre) data link system at the tips of the Bear-D's horizontal tail.

An F-4 escorts a visiting Tu-95RTs, keeping a respectful distance from the Bear. Interestingly, the Phantom is a US Air Force F-4E-36-MC (66-0328), not a US Navy F-4J or F-4S, indicating that the interception has taken place not far from the continental USA. According to Phantom pilots, the roar of the Tu-95's big turboprops could be heard in the cockpit while still more than a mile away!

◄ ◄ A 43rd TsBP i PLS Tu-95K coded «61 Red» brakes to a halt on the taxiway at Dyaghilevo airbase after a training sortie in 1991. The outer engines have been shut down during taxiing to save fuel. Note the blue-painted propeller spinners; coloured spinners or spinner tip are commonly used on Soviet/CIS Air Force turboprop aircraft to designate squadrons in the regiment operating the aircraft.

▲ Even before the big props have come to a standstill, technicians hook up a towbar to the nose gear of Tu-95K «61 Red» as a KrAZ-260 six-wheel drive truck moves into position, ready to push the aircraft back into its parking space visible beyond. The haste is understandable – regulations require the taxiway to be vacated as quickly as possible. Note the red anti-collision beacon between the automatic direction finder (ADF) strake aerials on the bomber's fuselage and the concrete block on the truck bed to give the tug more traction. «61 Red» wears a faded «Excellent aircraft» badge on both sides of the nose; the aircraft itself is very probably in excellent condition but the badge is not!

One more of the Centre's Bear-Bs takes off on a training flight. Due to the lack of IFR capability and hence their relatively short range the surviving Tu-95Ks were relegated to the training role in the 1980s.

◄

▲
Another shot of Tu-95K «61 Red» as it taxies in at Dyaghilevo AB in 1991.

The captain of a Tu-95 has opened the portside sliding direct-vision window to let in some fresh air into the flight deck, now that the mission is over. The flight deck roof of early versions of the Bear was rather low, which was a major grievance for the crews because of poor visibility, but this was remedied on later variants. Note the BTs-61 star tracker for celestial navigation and the reinforcement plate around the lateral observation blister.
◀

The crew of a Tu-95KM coded «64 Black» is lined up beside the aircraft as the co-pilot reports mission readiness to te captain before a sortie. The variance in the airmen's clothing is noteworthy; only two of the men wear leather jackets while the rest are clad in black coats made of cotton fabric. The vehicle in the background is a ZiL-131 truck equipped as a ground power unit.

Death row. To comply with arms reduction treaties a large number of early-model Bears was scrapped at Engels AB near Saratov in southern Russia which, apart from hosting two heavy bomber regiments, is an equipment disposal centre. Here, left to right, are Tu-95K «42 Red», Tu-95K-22 «32 Red», Tu-95K «62 Red», Tu-95K-22 «34 Yellow», Tu-95K «61 Yellow», Tu-95K-22s «24 Red» and «27 Red» and five more Bear-Gs, all with their fuselages cut into three pieces by a mechanical guillotine.

▲◀ «32 Red», a 43rd Combat and Conversion Training Centre Tu-95K, parked at Dyaghilevo AB in 1991, with several sister ships visible beyond. The tactical code is applied over a patch of fresher silver paint, clearly indicating the aircraft has been recoded.

▲ «34 Red», another Tu-95K, on the neighbouring hardstand. On this aircraft even the propellers are under wraps, indicating it is certainly not due to fly shortly, while on the next aircraft in the lineup the flight deck section is wrapped up. The quasi-civil Tu-154M airliner at the remote dispersal is used as a staff transport.

Another view of «32 Red», with traces of a recent snowstorm still visible on and around the aircraft. The engine nacelles are wrapped in tarpaulins and the star tracker protected by a red metal cover. Note the boarding ladder and the obligatory fire extingusher on a tripod parked in front of the aircraft.
◀

▲
«I will eat ya.» In life the Tu-95K was intimidating enough, but this stripped-out hulk sitting on the outskirts of Dyaghilevo AB in the spring of 1991 makes a positively hair-raising picture. The gaping hole where the radome and antenna of the YaD radar used to be looks like the gaping mouth of some Martian monster. The absence of the heavy radar set has caused the aircraft to assume a nose-high position. The blue-painted landing gear struts of this particular example are noteworthy.

▶
▶
These views of the same dead Bear-B show the nose's internal structure with attachment fittings for the radar's search (lower) and missile guidance (upper) antennas, as well as the weapons bay. Note also the Tu-22M3 bomber and Il'yushin IL-76MD transport in the background awaiting overhaul at the co-located Aircraft Repair Plant No. 360.

This view shows the Bear-B's large BD-206 missile rack in fully lowered position, as it would be immediately before missile release. The rack swings down on hefty arms to lower the Kh-20 cruise missile clear of the fuselage for engine starting. Note the fold-away restraining arms on the rack which keep the missile from rocking before release. The missile air intake fairing has «bled» down into the extended position, which of course is impossible in flight if the missile is lowered.

The Bear-B's six weapons bay doors fold inwards to accommodate the Kh-20 missile in semi-recessed position during take-off and cruise to the target. After missile launch they closed to fill the space around the the BD-206 missile rack which formed part of the lower fuselage skin in fully extended position. The missile air intake fairing is rotated into the stowed position in this view.

Another view of the weapons bay with the three portside doors open and the starboard doors closed. Note the hefty lock in the centre of the rack which engaged the missile's mounting lug; the fold-away restraining arms can be seen in retracted position fore and aft of it. This weapons bay design rendered the Tu-95K/Tu-95KM unsuitable for carrying free-fall bombs.

▲
The red waistband on this Tu-95K, «33 Red», basking in the April sun at Dyaghilevo AB identifies it as a «war weary» Tu-95KU trainer. Note the NK-12 engine lying on the grass beside the aircraft (on the left).

▶
The large hydraulic jacks standing beside Tu-95K «32 Red» indicate that the aircraft is due for a landing gear retraction check soon.

Tu-95K «61 Red» is already in wraps at the end of a flying training day, but maintenance work is still in progress, as indicated by the UPG-300 ground power unit on a ZiL-131 truck chassis parked beside the aircraft.

▶

RUSSIAN AIRCRAFT *IN ACTION*

33  TUPOLEV. TU-95

«Want a racing car?» With the outer wings and rear fuselage/tail unit gone but the inner engines and propellers still in place, these Bears (Tu-95K-22 «32 Red» and Tu-95K «42 Red») scrapped at Engels make an interesting sight. Theoretically, climb in, start the engines and... This view also illustrates the Bear-G's missile pylons.

More «racing cars» of the same type in the «boneyard» at Engels; note that the aircraft coded 32 Red is a different one, being a probeless Tu-95K! The nearest aircraft of the three, Tu-95K-22 «20 Red», carries five mission markers on the nose to signify... no, not the number of points won in races, of course, but successful live missile launches during exercises.

The No. 4 NK-12MV engine is salvaged from a retired Tu-95K-22, «55 Red» (note the rear ECM fairing just visible on the left). It will eventually keep other Bears flying for some time yet. In Soviet/Russian aeronautical slang such aircraft reduced to spares to keep others flyable are called «Avrora» – an allusion to the famous cruiser RNS Aurora which signalled the beginning of the 1917 October Revolution by firing her nose gun and was later permanently moored on the Neva River in Leningrad (now St. Petersburg) as a museum. Alas, unlike the cruiser, the only thing awaiting such «Auroras» is the breaker's torch...

Another view of the same aircraft sitting on the grass at Engels; the wheels are positioned on concrete slabs to prevent the aicraft from sinking into the mud when the rains come. «55 Red» was one of several Bear-Gs used for radiation reconnaissance duties; it still carries RR8311-100 air smapling pods under the outer wings.

▲
A Tu-95MS-6 (izdeliye VP-021; NATO reporting name Bear-H) coded «01 Red» takes off from runway 12 at Zhukovskiy. This was the final major production version of the Bear; it is normally armed with six Raduga Kh-55 (NATO code name AS-15 Kent) cruise missiles on a rotary launcher in the bomb bay.

«01 Red», which belongs to the Tupolev Design Bureau, is a weapons testbed of some sort, as revealed by the video camera fairing under the starboard wingtip just visible in the lower photo. Unusually, this example lacks the electronic support measures (ESM) fairings under the tail gunner's station.
▼

▲
Seen from the refuelling systems operator's station, a Tu-95MS-6 takes on fuel from an Il'yushin IL-78 tanker (NATO reporting name Midas) as the two aircraft fly high over a thick layer of overcast. The IL-78 is a three-point tanker but heavy aircraft always use the centre hose drum unit offset to port. This view illustrates the Bear-H's much neater «duck bill» radome housing an Obzor (Perspective) radar, as well as the raised flight deck roof offering more headroom.

«317 Black» (c/n 19317), another Tu-95MS-6 operated by the Tupolev OKB, begins its landing gear retraction sequence during a demonstration flight at MosAeroShow '92 in Zhukovskiy (11th-16th August 1992). The aircraft is likewise a weapons testbed; note the camera fairing under the starboard wing – a test equipment item whose design has not changed for decades.

Tu-95MS-6 «23 Black» completes its landing run on runway 30 at Zhukovskiy; the Nos. 1 and 4 engines have been shut down already. The aircraft, which presumably belongs to the 1096th GvTBAP (Guards Heavy Bomber Regiment) at Engels-2 AB, has obviously arrived for some kind of modification work. It is a late-production example, as indicated by the auxiliary power unit built into the fin root fillet; the open APU intake door and nozzle.

Seen from the control tower at Zhukovskiy, Tu-95MS-6 «317 Black» begins its take-off run from runway 12 in August 1992. Minutes later the aircraft came back to make a high-speed pass over the runway.

Apart from the standard Tu-95MS-6, the Bear-H existed in a more potent form – the Tu-95MS-16 which could carry sixteen Kh-55 cruise missiles instead of the usual six. Illustrated here is an example of this variant.

A Tu-95MS sits parked at the Russian Air Force Research Institute (GK NII VVS) airfield in Akhtoobinsk Oddly, the aircraft is coded «55 Red» on the tail and «56 Red» on the nose gear doors – probably as a result of a painting error in the latter case. On the Tu-95MS all dielectric parts and some cowling panels were painted white.

Two more views of an uncoded Tu-95MS-16 at the GK NII VVS airfield in Akhtoobinsk. These photos show the large pylons under the wing roots, each of which could carry six Kh-55 missiles, and the tandem hardpoints between the inner and outer engines (visible in the lower photo) for carrying two more Kh-55s on each side. The actual racks on which the missiles were suspended appear to have been removed. The Tu-95MS-16s were later downgraded to Tu-95MS-6 configuration in compliance with the START-1 strategic arms reduction treaty which limited the number of nuclear warheads to be carried by a single aircraft.

Tu-95MS-6 «31 Black», an example operated by GK NII VVS, wears an unusual bagde depicting a Russian knight; the signinficance of the badge is unknown. This view shows the radar warning receiver antennas on the sides of the nose below the IFR probe and the the hemispherical Mak-UFM missile warning system (MWS) sensor under the nose which are a feature of late-production Bear-Hs.

A Tu-95MS-6 streams contrails across the sky as it cruises along at high altitude. The aircraft looks nice and clean, except for the tell-tale exhaust stains.

A view of a Tu-95MS from an inspecting NATO fighter. Even in post-Soviet times the Bear-Hs keep NATO's air defence forces busy!

«33 Black», a late-production Tu-95MS-6 belonging to the 1096th GvTBAP at Engels-2 AB. An IL-78 tanker operated by the co-located 1230th APSZ (Aerial Refuelling Regiment) is just visible beyond; despite their obviously military role, many IL-78s wore civil registrations and the full livery of the Soviet airline Aeroflot.

▲
A Tu-95MS-6 formates with an IL-78 (wearing overt military markiings this time but no tactical code). The high ocation of the centre hose drum unit shows this is a convertible IL-78 tanker/transport, not an IL-78M which is a dedicated tanker variant.

The underside of a red-coded Tu-95MS-6. Note the bulged nose-wheel well doors and the long conduit running along the port side from the IFR probe to the rear fuselage fuel tanks.
◄

RRREEEAAARRR... As one of the highlights of the flying display at the MAKS-93 airshow in Zhukovskiy (31st August – 5th September 1993), Tu-95MS-6 «317 Black» performs a high-speed run for the benefit of the crowds – and the sound of it was just as impressive as the sight. Note the black photo calibration markings on the starboard side of the rear fuselage and what looks like a camera mounting on the side of the fin, both indicative of the aircraft's trials status.

The demo flight completed, «317 Black» rolls along runway 30 in Zhukovskiy. Again the outer engines have been shut down even before the aircraft has left the runway; this appears to be standard operational procedure.

◄◄
«36 Black» (c/n 35793), a late-production Tu-95MS-6 from Engels, is manoeuvred into position by an APA-5D ground power unit at Kubinka AB west of Moscow on 12th April 1992. The APA-5D based on a Ural-375D six-wheel drive truck chassis is often used as a tug, albeit usually not for such heavy aircraft. A Soviet Air Force/Antonov Airlines An-124 Ruslan heavy transport («08 Black»/CCCP-82038) is just visible beyond.

▲
Tu-95MS-6 «17 Black» formates with IL-78M «35 Blue» over Kubinka AB on 7th August 1997 in a simulated refuelling procedure during the dress rehearsal for the 16th Air Army's gold jubilee celebrations which took place the following day.

Flanked by two non-airworthy Tu-160 bombers, Tu-95MS-6 «26 Red» (c/n 32477) parked at the Tupolev OKB's flight test facility in Zhukovskiy during the MAKS-95 airshow (22nd-27th August 1995).
▼

Another view of «36 Black» on the demonstration facility apron at Kubinka AB during the base's first open doors day on 12th April 1992, with a Tu-22M3 parked beyond. The aircraft is one of two 1096th GvTBAP Tu-95MS-6s which later paid a courtesy visit to Barksdale AFB, Georgia, in 1994. The 1992 event was messed up a bit by fickle weather, with sudden snowstorms every now and then, as is evident on the photo.
◄

As illustrated here, Tu-142s routinely shared bases with their «close relatives who were also in the Navy» – Tu-95RTs OTH targeting/maritime reconnaissance aircraft.

«20 Red», a Tu-142M (izdeliye VPM; NATO reporting name Bear-F Mod 2) long-range anti-submarine warfare aircraft at Kipelovo AB. Since the resident North Fleet Bear-Fs have black tactical codes, this one may be a Pacific Fleet example from OokraXnka AB arriving for a visit to the local maintenance base. The Tu-142M, which was the third version of the Bear-F, introduced a 0.3-m (1-ft) forward fuselage stretch and a redesigned flight deck to improve the crew's working conditions; the raised flight deck roof was later incorporated into the Tu-95MS which is basedon the Tu-142M airframe.

A Soviet Navy crew is briefed in front of their ship, a Tu-142M or Tu-142MK, sometime in the 1980s. The difference in clothing style is noteworthy. The cloth «handbags» are used for carrying flying helmets.

▲
This uncoded example of the Tu-142 (izdeliye VP/Bear-F) ASW aircraft is unique in that it features an aft-pointing dielectric fin tip fairing for a magnetic anomaly detector (MAD). This view shows clearly the 12-wheel main gear bogies of the initial production version with three rows of four wheels – a feature intended for soft-field operations which quickly earned the Tu-142 «sans suffixe» the nickname «Sorokonozhka» (Centipede). Note also the chin fairing housing a thermal imager, the low flight deck roof (both characteristic of the «centipede») and the bulged nose gear doors. The teardrop radome under the centre fuselage houses a Berkoot (Golden eagle) search radar.

▶
The latest and most advanced ASW version of the Bear is the Tu-142MZ (izdeliye VPMZ/Bear F Mod 4). This mostly natural metal example coded «56 Black» belongs to the 240th GvOSAP (Guards Independent Composite Air Regiment), i.e., the Russian Navy' Combat and Conversion Training Centre at Ostrov AB near Pskov. Apart from the fin tip MAD, the Tu-142MZ can be identified by the shallow chin fairing and the small «thimble» at the tip of the navigator's station glazing, both associated with the Zarechye sonar system (hence the letter Z in the designation).

Routine maintenance underway on a pair of 240th GvOSAP Tu-142MKs, «90 Black» and «93 Black», at Ostrov AB.

▲
Tu-142MK «93 Black» (c/n 1603062) taxies out for a training sortie at Ostrov past a line of sister ships and Il'yushin IL-38 shore-based ASW aircraft.

A naval aircraft is not what you would normally expect to see during the annual VE-Day and Aviation Day flypasts at Moscow-Tushino. The reason why this Tu-142MZ (escorted by two equally naval Sukhoi Su-27K fighters of the 279th Shipboard Fighter Regiment) appeared over Tushino in 1996 is that the Russian Navy was celebrating its 300th anniversary that year.
◄

Another view of the Tu-142MZ with Flanker-D escort as it makes a pass over Moscow-Tushino in 1996. Although the main anniversary festivities (including a big flying display) took place in St. Petersburg which is the birthplace of he Russian Navy, the Naval Air Arm sent this «delegation» to the capital.

Since the requirement to operate from dirt strips was soon abandoned as unwarranted, the 12-wheel main gear bogies of the initial production Tu-142 were replaced by four-wheel bogies from the second version onwards (which, oddly enough, was also designated simply Tu-142 with no suffix letters); this provided a substantial weight saving. Illustrated here is a Tu-142MK.

# RUSSIAN AIRCRAFT IN ACTION

TUPOLEV. TU-95

◄ The Tu-142MK (izdeliye VPMK, or Bear-F Mod 3) introduced a more capable Korshoon (Kite, the bird) search radar and an MMS-106 Ladoga MAD at the tip of the fin. Unlike the modified Tu-142 «sans suffixe» depicted on page 52, the MAD fairing was angled slightly upward. Illustrated here is «90 Black» which, like the other Bear-Fs at Ostrov, is maintained in excellent condition. The streakjed appearance of the propeller blades is noteworthy. Note also the non-standard access ladder and the ground-to-crew telephone connected to a socket on the forward fuselage.

▲ Ostrov AB is a well-built and well-equipped base which makes a favourable contrast with some Russian airbases. The Tu-142 hardstand is a good example, this view illustrating the high-quality surface and the solid blast deflector at the far edge paved with concrete slabs. It is designed to withstand the jet blast of Tu-22M jet bombers which also operate from Ostrov.

Maintenance underway on Tu-142MK «93 Black» prior to a training flight, with a UPG-300 starter cart providing electric power. Note that all four engines have their cowlings open. The vehicle on the right is a GAZ-66 cabover, a very common four-wheel drive army truck. ◄

◀ Tu-142MK «93 Black» (c/n 1603062) on short finals to its home base. This aircraft took part in the Royal International Air Tattoo '94 at RAF Fairford.

A Tu-142MK passes overhead, showing off the white-painted engine nacelle bottoms, the white radome – and the unspeakably sooty wing undersurface aft of the jetpipes. The slender pencil-shaped fuselage and the high aspect ratio sharply swept wings give the Bear a racy look.
▼◀

▲
Tu-142MK «80 Black» is in a temporary state of disrepair, missilng the propeller on the No. 3 engine; the engine itself is covered with a tarpaulin.

Another view of Tu-142MK «90 Black». Note the dielectric dome covering a satellite navigation antenna on the forward fuselage.
▼

«95 Red», an operational North Fleet Tu-142MK operated by the 76th OPLAP DD (Independent Long-Range ASW Air Regiment), lands at Kipelovo AB in the autumn of 1998. The eight large mainwheel tires furnish an impressive puff of smoke as the aircraft «burns rubber» at the moment of touchdown.

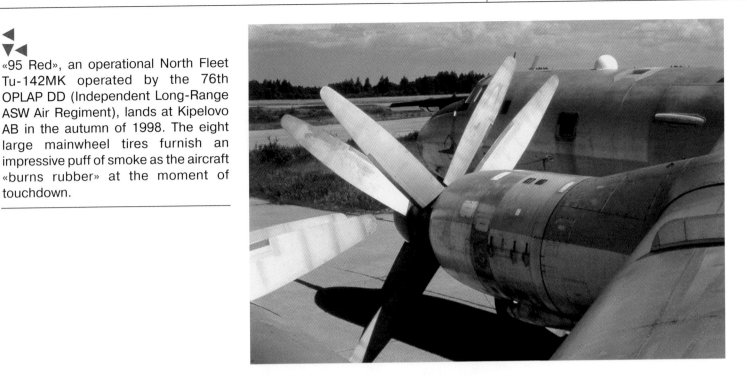

A view of the No. 2 Kuznetsov NK-12MV turboprop and AV-60N contra-rotating propellers of a Tu-142MK. Note also the dorsal escape hatch used in the event of a ditching.

The long-range navigation (LORAN) strake aerial offset to port and the SAT-NAV antenna dome of a Tu-142MK.

Another view of the same aircraft as it accelerates along the runway at Kipelovo, taking off on another sortie in a green and gold autumn setting.

A full frontal of a Tu-142MK.

76th OPLAP DD Tu-142MKs lined up at Kipelovo AB. «51 Black» in the foreground wears an «Excellent aircraft» badge. Most Bear-Fs had a natural metal finish.

The flight line at Ostrov AB with Tu-142MK «90 Black» in the foreground. Just discernible in the background on the right are Tu-154M RA-85609 used as a personnel shuttle and IL-20RT CCCP-75481, a former space tracker aircraft now used as a trainer for the IL-38 crew.

The Tu-142MK can be distinguished from the later Tu-142MZ (with which it shares the tail «stinger») by the smooth nose contour.

▲
As this view of a Tu-142MK on final approach shows, the main gear bogies assume a slightly nose-down position in weight-off condition.

▶
▶
Even keeping an aircraft in good enough condition to earn the «Excellent aircraft» badge is by no means a guarantee against failures, as this unserviceable Tu-142MK («80 Black») illustrates. The GSh-23 cannons have been removed from the UKU-9K-502 tail turret.

An Indian Navy Tu-142MK-E serialled IN 314 seen at its birthplace, the Taganrog Machinery Plant named after Gheorgiy Dimitrov, where it underwent repairs in 1995. The ARK tail code referes to the aircraft's home base, Indian Navy Air Station (INAS) Archelon; other examples based at INAS Dabolim, wear the tail code DAB.

Close-up of the squadron badge on the nose of Tu-142MK-E IN 314. India was the sole export customer for the Bear, taking delivery of eight Tu-142MK-Es.

An uncoded Tu-142MZ (possibly c/n 2605426) takes off from rain-soaked runway 12 at Zhukovskiy during the MAKS-93 airshow. This and several other Bear-F Mod 4s were painted grey overall; the significance of the white areas on the nose is unknown.

The same aircraft completes its landing run after the demo flight.

The Tu-142MR (izdeliye VPMR) was a derivative of the Tu-142MK designed to provide communications between land-based or airborne command posts and submerged missile submarines – the Soviet counterpart of the US Navy's Boeing E-6A TACAMO (TAke Charge And Move Out). Here, «11 Black», the Tu-142MR prototype, is seen at the Ukrainian Navy's 33rd Combat and Conversion Training Centre at Nikolayev-Kul'bakino AB

Unlike production examples, the prototype was converted from a Tu-142MK airframe and retained the Bear-F's glazed nose, with the forward antenna in a chin fairing. It is seen here in somewhat «unbuttoned» for maintenance at the Russian Navy's Kipelovo airbase.

RUSSIAN AIRCRAFT *IN ACTION*

◀
◀
«15 Black», an operational North Fleet Tu-142MR, lands at Kipelovo AB, its home base, in the autumn of 1998.

▲
«12 Black», the first production Tu-142MR, runs up the Nos. 1 and 2 engines at Kipelovo in the autumn of 1998. Note the ground power cables connected to sockets aft of the nose gear unit.

Close-up of the massive trailing wire aerial (TWA) drum fairing of a production Tu-142MR. The greater part of the fairing is dielectric. Note the stabilising drogue (with perforations around the perimeter) semi-recessed in the fairing and the air intake of the ram air turbine driving the TWA drum; the air exits through the ventral aperture just visible ahead of the dielectric portion. The technician provides scale, illustrating the sheer size of the thing. ▶

TUPOLEV. TU-95

◀ This view of «12 Black» illustrates the characteristic nose treatment of production Tu-142MRs. The nose glazing was deleted and the undernose antenna was moved to a thimble fairing in the extreme nose to give it a better field of view. The shallow chin fairing is similar to that of the Tu-142MZ.

▲ Production Tu-142MRs on a rain-soaked ramp at Kipelovo AB in 1998. The Bear-Js share the flight line with 76th OPLAP DD Tu-142MKs. ▼

◀ All production Bear-Js save one were painted medium grey overall. «15 Black» appears to have a slightly different undernose antenna fit.

▲ A view of the Tu-142MR's tail surmounted by the characteristic forward-pointing HF probe aerial. Interestingly, the prototype lacked the missile warning system aerials under the tail genner's station.

Nearly half of Tu-142MR's production run is illustrated in this view. Interestingly, the example farthest from the camera appears to be coded «21 Red».
▼◀

Another view of the first production example engine running at Kipelovo. Like the prototype, it has a natural metal finish.
▼

▲ The star insignia on the port side of Tu-142MR «16 Black» has been almost obliterated by the elements. The aerial under the chin fairing is enclosed by a red protective cover, while the white paint has chipped away from the ESM fairings on the sides of the nose, revealing the brown glassfibre.

▶▲ Front view of a Tu-142MR. The white «thimble nose» creates a rather strange impression.

Another view of Tu-142MR «16 Black». Interestingly, the star insignia on the starboard side is perfectly legible.

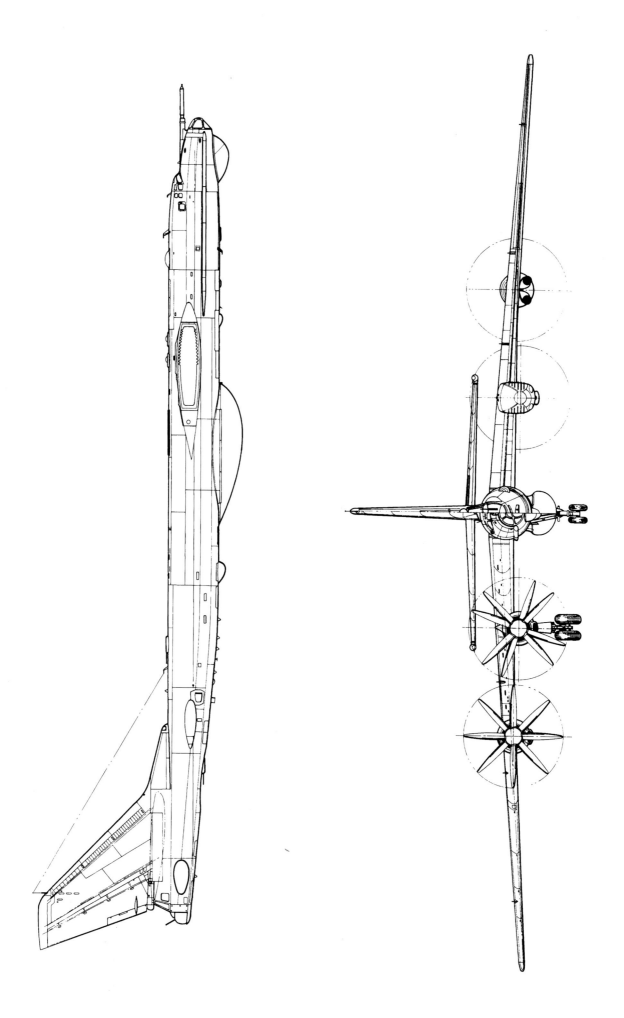

Starboard side view of the Tu-95RTs Bear-D (The wing is omitted for clarity).

Front view (left half of the picture) and rear view of the Tu-95RTs.

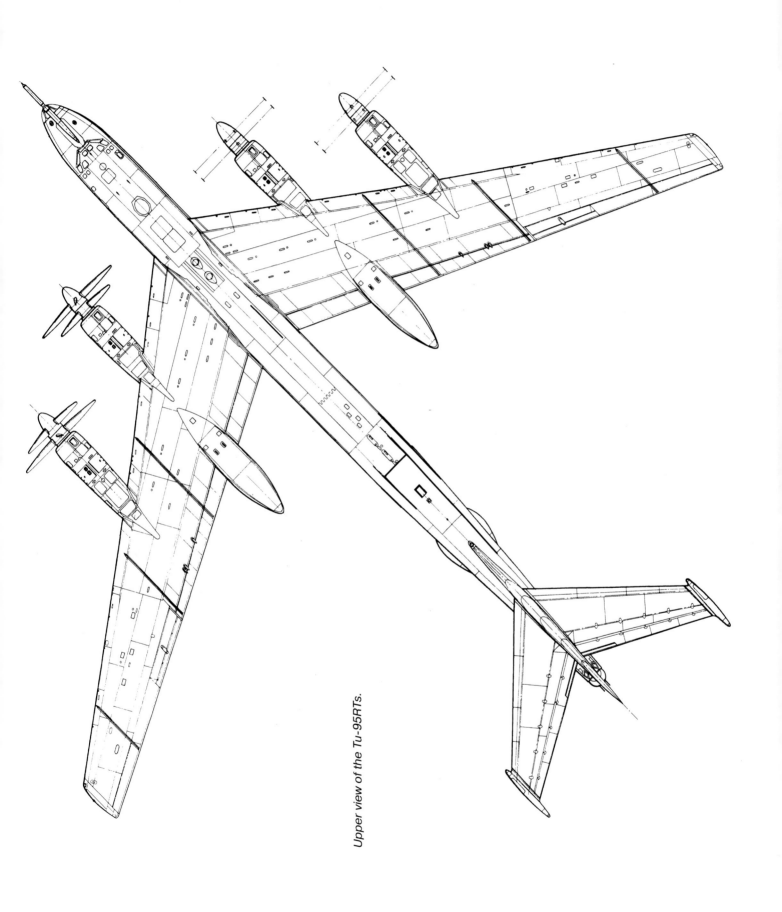

*Upper view of the Tu-95RTs.*

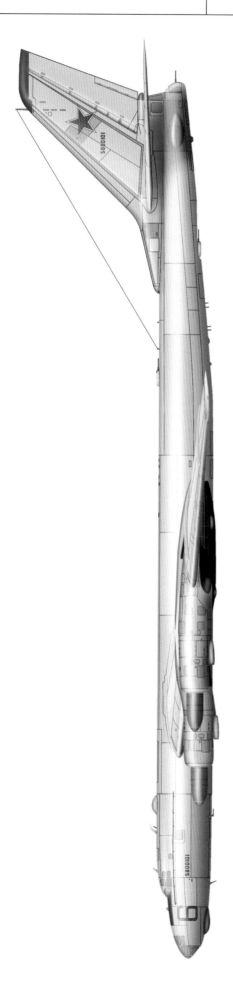

*The first production Tu-95 Bear-A (c/n 5800101).*

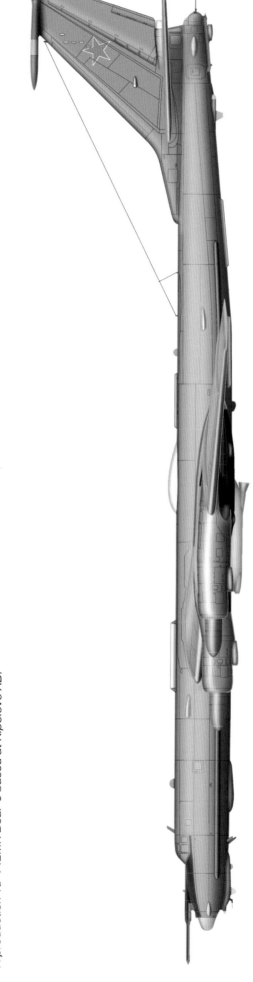

*A production Tu-142MR Bear-J based at Kipelovo AB.*